W9-AXC-910

Our Universe

# Jupiter

by Margaret J. Goldstein

Lerner Publications Company • Minneapolis

Lerner Publications Company
A division of Lerner Publishing Group
241 First Avenue North
Minneapolis, MN 55401 USA

Website address: www.lernerbooks.com

Words in **bold type** are explained in a glossary on page 30.

Library of Congress Cataloging-in-Publication Data

Goldstein, Margaret J.
    Jupiter / by Margaret J. Goldstein.
      p.   cm. — (Our universe)
    Includes index.
    Summary: An introduction to Jupiter, describing its place in the solar system, its physical characteristics, its movement in space, and other facts about this planet.
    ISBN: 0-8225-4652-3 (lib. bdg. : alk. paper)
      1. Jupiter (Planet)—Juvenile literature. [1. Jupiter (Planet)]
I. Title. II. Series.
QB661 .G66 2003
523.45—dc21                    2002000948

Manufactured in the United States of America
1 2 3 4 5 6 — JR — 08 07 06 05 04 03

This giant planet is covered with colorful, swirling clouds. It has rings and many moons. What planet is it?

**Earth**

**Jupiter**

The planet is Jupiter. Jupiter is much larger than our planet Earth. More than 1,000 planets the size of Earth could fit inside Jupiter.

Jupiter is the largest planet in the **solar system.** The solar system has nine planets. Earth is one of the nine planets.

All of the planets in the solar system **orbit** the Sun. To orbit the Sun means to travel around it. Jupiter is the fifth planet from the Sun.

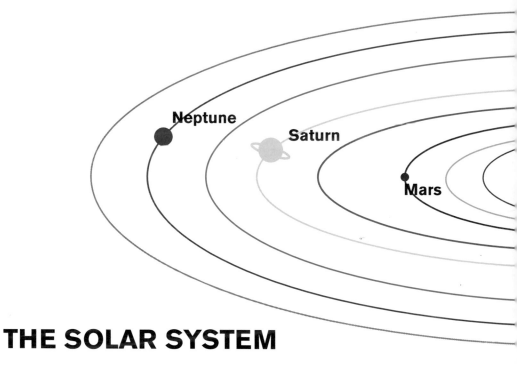

**THE SOLAR SYSTEM**

Jupiter makes a long trip around the Sun. It takes almost 12 years to orbit once. Earth orbits the Sun in only 1 year.

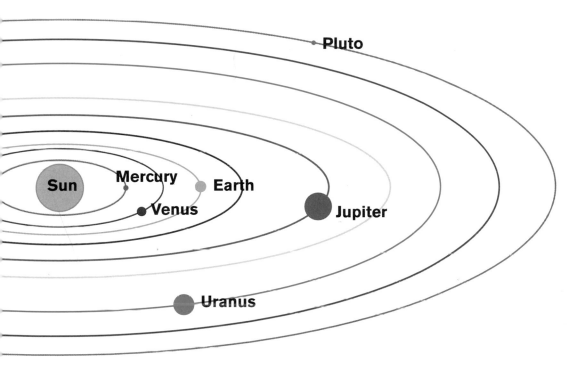

Jupiter **rotates,** too.  To rotate means to spin around like a top.  Jupiter rotates faster than any of the other planets in the solar system.  It rotates all the way around in about 10 hours.  Our planet takes 24 hours to rotate all the way around once.

Jupiter is not a solid planet like Earth is. It does not have hard ground. The planet is surrounded by a thick layer of gases. This layer is called an **atmosphere.** A hot, soupy layer lies below the atmosphere. Jupiter's center may be either liquid or solid.

# JUPITER'S LAYERS

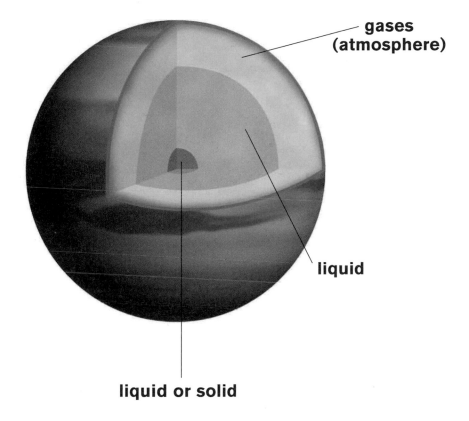

gases
(atmosphere)

liquid

liquid or solid

Jupiter's atmosphere is very windy. The winds blow colorful clouds around the planet. The clouds are red, brown, blue, yellow, and white. They look like stripes and swirls.

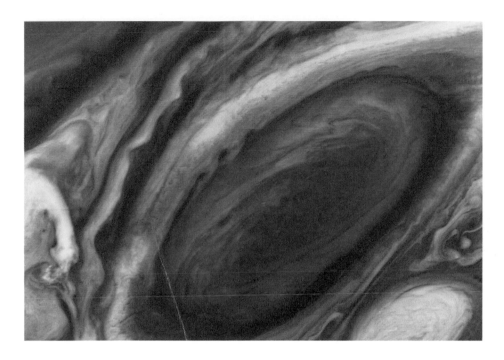

Jupiter's atmosphere is also stormy. The biggest storm is called the Great Red Spot. This giant storm cloud looks like a big red oval. It is so big that three Earths could fit across it!

Jupiter does not get much warmth from the Sun.  So the outside of Jupiter is very cold.

But Jupiter grows hotter below the atmosphere.  The center of the planet is probably burning hot.

Jupiter has three narrow rings. The rings orbit Jupiter. They are very hard to see from space. The rings are made mostly of dust. What else orbits Jupiter?

If we could see Jupiter's rings, they would look very thin.

At least 39 moons orbit Jupiter. Some of the moons are very small. Other moons are large. The four largest moons are named Ganymede, Callisto, Europa, and Io.

Ganymede is the largest moon in the solar system. It is much larger than Earth's moon. It is even larger than the planet Mercury.

Ganymede is made of rock and ice. The ground is covered with deep holes called **craters.** Ganymede also has tall mountains and deep valleys.

Many craters also cover Callisto. This moon has more craters than any other moon in the solar system.

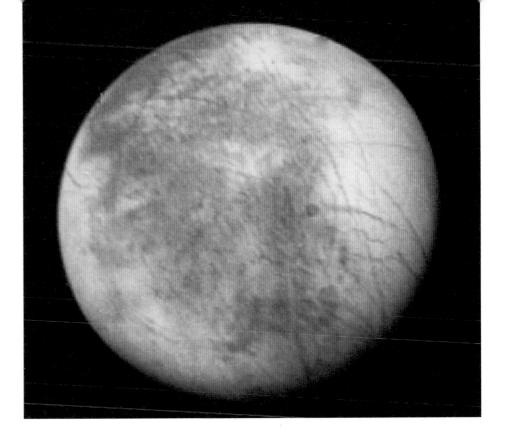

Europa does not have a lot of craters. It has a smooth icy covering with many streaks. The covering looks like a cracked eggshell.

Io is a red, orange, and yellow moon. It looks a little like a cheese pizza. The colors come from **volcanoes** on the moon.

Volcanoes are places where melted rock and hot gases burst out of openings in the ground. Io has more volcanoes than any other place in the solar system.

Jupiter looks like a bright star in our night sky. With a telescope, it is easier to see. But people wanted to see the planet up close.

In 1972, Americans sent a spacecraft to Jupiter for the first time. The spacecraft was called *Pioneer 10*. Six spacecraft have traveled to the planet since then.

These spacecraft tested Jupiter's atmosphere. They took pictures of the clouds and the Great Red Spot. They discovered Jupiter's rings and some of its moons.

Another spacecraft may travel to Jupiter
in the future. What would you do if you
could travel on the spacecraft? What
do you think you would find on Jupiter?

# Facts about Jupiter

- Jupiter is 484,000,000 miles (778,000,000 km) from the Sun.

- Jupiter's diameter (distance across) is 88,700 miles (143,000 km).

- Jupiter orbits the Sun in 12 years.

- Jupiter rotates in 10 hours.

- The average temperature in Jupiter's atmosphere is −160°F (−110°C).

- Jupiter's atmosphere is made of hydrogen and helium.

- Jupiter has at least 39 moons.

- Jupiter has 3 rings.

- Jupiter was named after the Roman king of the gods.

- Jupiter has been visited by *Pioneer 10* in 1973, *Pioneer 11* in 1974, *Voyager 1* in 1979, *Voyager 2* in 1979, *Ulysses* in 1992, *Galileo* in 1995–1999, and *Cassini* in 2000.

- Jupiter's large moons are called Galilean moons. They are named for astronomer Galileo Galilei. He discovered the moons about 500 years ago.

- Jupiter got its name because it is the largest planet in the solar system.

- Jupiter is also the heaviest planet in the solar system. It weighs more than twice as much as all the other planets combined.

- Jupiter's rings were discovered by the *Voyager 1* spacecraft in 1979.

# Glossary

**atmosphere:** the layer of gases that surrounds a planet or moon

**craters:** large holes on a planet or moon

**orbit:** to travel around a larger body in space

**rotate:** to spin around in space

**solar system:** the Sun and the planets, moons, and other objects that travel around it

**volcano:** an opening in the surface of a moon or planet. Hot rock and gases burst up through the opening.

# Learn More about Jupiter

## Books

Brimner, Larry Dane. *Jupiter.* New York: Children's Press, 1999.

Simon, Seymour. *Jupiter.* New York: Morrow, 1988.

## Websites

Solar System Exploration: Jupiter
< *http://solarsystem.nasa.gov/features/planets/jupiter/jupiter.html*>
Detailed information from the National Aeronautics and Space
Administration (NASA) about Jupiter, with good links to other help-
ful websites.

The Space Place
<*http://spaceplace.jpl.nasa.gov*>
An astronomy website for kids developed by NASA's Jet
Propulsion Laboratory.

StarChild
<*http://starchild.gsfc.nasa.gov/docs/StarChild/StarChild.html*>
An online learning center for young astronomers, sponsored by
NASA.

# Index